I0755962

PHILOSOPHY AND PHILOSOPHERS

Great Thinkers
And the Way They Think

THEODORE H. KITTELL

GlobalEdAdvancePress™
37321-7635 USA
www.GlobalEdAdvance.org

Philosophy and Philosophers
Great Thinkers and the Way They Think

Library of Congress Control Number:2009921426
ISBN 978-1-935434-03-0

Kittell, Theodore H., 1934 –

Subject Codes and Description:
1 SCI075000: Science: Philosophy & Social Aspects;
2 PHI000000: Philosophy: General;
3. BIO000000: Biography &Autobiography: Philosophers

Cover design by Barton Green

Printed in the United States of America

Published by
GlobalEdAdvancePress
37321-7635 USA

About the Author

Distinguished Professor, Theodore H. Kittell, PhD, JD, DLitt, has taught graduate courses in Organizational Behavior at Oxford Graduate School (1990-2005). He has two earned doctorates: PhD in Organizational Management from Walden University and JD from William Mitchell School of Law, and holds the DLitt from OGS. He earned the Master of Hospital Administration from the University of Minnesota and a Bachelor of Business Administration from the University of New Mexico. He has been a consultant for the Tennessee Board of Health.

Admitted to the Bar in California and Minnesota, Dr. Kittell is an expert witness in healthcare management and evaluates medical malpractice cases. He is a licensed Tennessee Nursing Home Administrator. During retirement he has pursued advanced education earning additional doctorates in Business Administration and Christian Psychology. He has served on numerous health-related boards. Throughout his business careers he has been active in Lions or Rotary service clubs.

Dr. Kittell has a lifetime interest in philosophy and has been active in short-term funded overseas seminars in Continuing Legal Education, Healthcare Management, and Men's Studies, and does mediation and arbitration work. He has done additional graduate study in Management at Harvard Graduate School of Business.

Dr. Kittell's experience includes 27 years as a hospital executive in the U.S. Air Force; Pulaski Memorial Hospital, Winamac, IN; Apple River Hospital, Amery, WI, and Coalinga District Hospital, Coalinga, CA. He has two decades of experience as a lawyer and nine years experience as a consultant in healthcare management. He has written and presented numerous papers in healthcare management, men's studies, assisted suicide and the right to die.

Dr. Kittell is a Diplomate in the Oxford Society of Scholars; a Life Fellow, American College of Healthcare Executives; Diplomate, Board of American Forensic Examiners; Fellow, Academy of Medical Administrators; Member, American College of Healthcare Administrators; Member, American Society of Healthcare Law, Medicine and Ethics; Member, Academy of Legal Studies in Business.

This little book is dedicated to

All readers who seek to understand
Philosophic ideas and their importance
Or the lack thereof,

And particularly dedicated to

My students over the years
Who have had difficulty
Understanding philosophy.

I also dedicate this book to

Paulette Armstrong
and thank her for the assistance
and encouragement she gave
me on the first draft.
Without her I was ready
to abandon this book.

Table of Contents

About the Author		5
Introduction		11
Definitions		13
Basics of Philosophy		19
Chapter One:	Seventh Century B.C. to the End of the Pre-Socratics	21
Chapter Two:	Socrates to Aristotle	29
Chapter Three:	Aristotle to the Dark Ages	33
Chapter Four:	The Dark Ages	43
Chapter Five:	The Post Dark Ages	49
Chapter Six:	European Philosophy	55
Chapter Seven:	20th Century Philosophers	63
Philosophical Dilemmas		81
Philosophers Used in This text		85
Sources		91

Introduction

There is no one definition of philosophy. To try to determine what philosophy is or to try and explain it is difficult. This is very evident from the two dictionaries listed below. There is not even agreement between the two dictionaries.

Philosophy is perhaps not so much a subject as it is an activity in which one would engage. Some authors define philosophy as conceptual analysis. One simple approach is to define it as thinking about thinking.

Philosophy to the casual observer will probably seem to be very complex. So much so you may not have a clue about what the philosopher may be talking.

Philosophers have developed their own jargon and what they say and write may seem complex. Many professors develop their own jargon, too. Jargon is not understood by people outside the academic field or profession.

Hopefully this book will assist your understand of some of the philosopher's jargon. This book is designed to help students better understand the subject of philosophy and philosophers. It has been designed to be used as a study

Before getting into trying to understand what various philosophers are talking and writing about, let us look at dictionaries various definitions of philosophy and philosophers. Philosophy is difficult to explain as is evident from the definitions given by the two dictionaries listed below. There is little or no agreement between the two dictionaries.

This book is about understanding philosophers and their philosophic ideas to help students understand what the subject of philosophy is all about. This book is to help them know whether the ideas expounded by a given philosopher was important or not important at all. The text covers philosophers from the 7th Century B. C. to the 20th Century A. D. The book also presents philosophic dilemma which students after reading this book will be able to take either the pro or con position regarding the dilemma.

Definitions

Philosophy & Philosopher

The easiest definition for philosophy and philosophers is those who think about thinking.

Merriam Webster's Collegiate Dictionary

Philosophy

1. All learning exclusive of technical precepts and practical arts.
2. The science and liberal arts exclusive of medicine, law and theology.
3. The pursuit of wisdom.
4. A search for a general understanding values and reality by speculative rather than observational means.
5. A system of philosophical concepts.
6. The most basic beliefs, concepts and attitudes of
7. Calmness, temper and judgment of a philosopher

Philosopher

- A person who seeks wisdom or enlightenment.
- A scholar or thinker.
- A student of philosophy.

- A person whose philosophical perspective makes understanding with equanimity easier.
- An expounder of a theory in a particular area of expertise.
- One who philosophizes.

American Heritage Dictionary

Philosophy

1. Speculative inquiry concerning the source and nature of human knowledge.
2. A system of ideas based on such thinking.
3. The sciences and liberal arts except medicine, law and theology.
4. The system of motivating values, concepts or principles of an individual, group or culture.
5. A basic theory concerning a particular subject.

Philosopher

- A specialist in philosophy.
- One who lives by a particular philosophy.
- One who takes a calm and rational approach to life.

The study of philosophy has a long history going back to the 7th century B. C. This book will deal with the Western concepts of philosophy. The Eastern concepts of philosophy deal with meditation and the various religions of India, China, and Japan.

All the important questions of philosophy are often asked by four-year old children.

Basics Of Philosophy

One of the basics is Ontology, the study of what there is. Another basic is Epistemology, the study of how we came to know about it. Epistemology is often linked to the names of individual philosophers. For example, "Plato's epistemology," or "Kant's epistemology."

Ontology is less frequently ascribed to individuals. The more developed things are, the more things there are supposed to be.

Willard van Orman Quine once stated that all the important questions of philosophy were often asked by four-year old children. The questions are:

"What is there ?" (ontology);

"How do you know ?" (epistemology);

"Why should I ?" (This might be construed as a question in ethics).

The most basic question children ask is **"Why."**

Chapter One

Seventh Century B.C. to The End Of the Pre-Socratics

Many of the early philosophers dealt with basic concepts of earth, air, fire, and water.

Philosophers and philosophy began in the 7th century B.C. with the Greek philosophers. The Greek philosophers of this early period are known as the "Pre-socratics." This label is misleading since it includes philosophers who came both before and after Socrates.

No one knows when the study of philosophy started; however, many of the early philosophers dealt with basic concepts of earth, air, fire, and water.

Herodotus (484 - 425 B.C.)

Herodotus reported the Egyptians were the first to assert the doctrine that the soul of man is immortal.

A Greek researcher, he was the world's first historian.

Herodotus' remarkable writings contained excellent ethnographic descriptions of the peoples that the Persians conquered, fairy tales, gossip, legends, and a humanitarian morale.

Thales of Miletus (c. 620-550 B.C.) was the first person to be recognized as a philosopher. Aristotle was the major source for Thales's philosophy and science. He identified Thales as the first person to investigate the basic principles, the question of the originating substances of matter and, therefore, as the founder of the school of natural philosophy.

Thales was interested in almost everything, investigating most areas of knowledge, philosophy, history, science, mathematics, engineering, geography, and politics. He proposed theories to explain many of the events of nature, the primary substance, the support of the earth, and the cause of change.

Thales's hypotheses were new and bold, and in freeing phenomena from religious intervention, he paved the way towards scientific investigation. His philosophy for which he is best known is that the earth floats on water, and that all things come to be from water. Also, he felt that even plants had "souls."

Anaximander (c. 610-550 B.C.) was one of the earliest Greek thinkers. According to available documents, he was the first philosopher known to have written down his studies. He was an early proponent of science and attempted to observe and explain different aspects of the universe, with a particular interest in its origins, claiming that nature is ruled by laws, just like human societies. With his assertion that physical forces, rather than supernatural means, create order in the universe, Anaximander is known to have conducted the earliest recorded scientific experiments. He proposed that everything was made of the boundary and was unlimited. Some think what he proposed was without meaning or value.

Anaximenes (c. 570-510 B.C.) stated that everything was made of air. He held that air was the source of all that exists at different degrees of density, and under the influence of heat and cold, which expands and con-tracts its volume. He believed air came in threads. "Just as our soul, being air holds us together, so do breath and air encompass the whole world."

Heracltus (c. 540-590)

Heraclitus put forth the idea that everything was made of fire. He also claimed that everything was in a state of flux. He stated that you could not step into the same river twice and there was no difference between up and down. You can identify this position as 'Heraclitus Metaphysic' meaning the flux. He is best known for his controversial teachings that things are constantly changing, that opposites coincide, and that fire is the basic material of the world.

Pythagoras (c. 570-10)

Pythagoras is known for inventing the right angle triangle. He also believed that everything is made of numbers. He was a proponent of reincarnation. He held that everything had souls, including plants and animals. This interfered with his diet and was indirectly responsible for his death.

Parmenides (c. 520-430)

Parmenides and **Melissus** (c. 480-420) did not claim that everything was made of one substance. They held that everything was in reality one thing, whether large, spherical, motionless, infinite, or changeless. All appearances of these occurring are an illusion. This theory was known as Monism, from the Greek word 'mono' meaning an antiquated recording.

Empedocles (c. 500-430)

Empedocies held that everything was made of earth, air, fire and water, held together by or broken down, by love and strife. These two were on an endless cycle of recurrences.

Zeno (c. 500-440)

Zeno was their successor and had a theory of a series of paradoxical statements that nothing can move. His arguments questioned whether space and time are infinitely divisible or whether one or both are made of indivisible quanta.

The last of the Pre-socratics were the Atomists. They often are given the distinction of holding forth a theory that atoms are the smallest things.

Democritus (c. 450-360)

Democritus and **Leucippus** (c. 450-390) both held that the atom couldnot be split. In modern timesit is known the atom can besplit.

Chapter Two

Socrates to Aristotle

Plato was a student of Socrates;
Aristotle was a student of Plato.

Socrates (c. 469-399)

Socrates did not write anything. Much of what we know that he said has come to us through the writing of Plato We really do not know what to attribute to Socrates and what to attribute to Plato.

Plato (c. 427-347)

Plato believed that objects such as tables and chairs are copies of perfect originals laid up in heaven appreciated by the so called Forms. These Forms have abstract ideas such as truth, beauty, goodness and love. The concept of Forms was challenged. If every thing is a copy of a good Form, there must be Forms of awful things. Plato listed hair, mud and filth.

Another question about **Plato** is (1) was he a feminist. or (2) not a feminist? Both points can be supported from his writings. The support for (1) is in Book 3 of the *"Republic"*; he stated that women should not be discriminated against in matters of employment because they are women. His position in (2) is also in Book 3 of the "*Republic*", where he stated that women are so much less talented than men by nature.

Plato taught that the soul is divine, uncreated, and immortal and passes through many incarnations. He did not elaborate much further on the subject of incarnations.

Pythagoras and many Greek philosophers studied in Egyptian colleges where they mastered the mystical arts and absorbed occult beliefs.

Chapter Three

Aristotle To The Dark Ages

"By pleasure we mean the absence of mental and physical pain."

The Epicureans

Aristotle (c. 382 - 322

Aristotle was also known as Stagirite. He was called by this name because he was from Stagira in Macedonia. He expected to succeed Plato as the head of the Academy. Instead **Speusippus** was made the head of the Academy. This made Aristotle very angry so he founded his own school, the Lyceum.

Aristotle was very intelligent. He made many contributions to Logic. In fact, he was the person who developed logic. He was instrumental in developing the Philosophy of Science as well as Biological Taxonomy, Ethics, Political Philosophy, Semantics Aesthetics, Theory of Rhetoric, Cosmology, Meteorology, Dynamics, Hydrostatics, Theory of Mathematics, and Home Economics. Some of these areas of philosophy have been developed further by other philosophers.

Some of his facts could be questioned. An example is from his writings on biology concerning his description of a snake's genitalia. From "De Generatione Anilmalium" he stated, "Snakes have no penis, because they have no legs, and they have no testicles since they are so long."

Aristotle's philosophy became fragmented and several rival schools were formed to complement and to argue with the Academy and Lyceum. The major new philosophy schools that arose at the beginning of the 3rd Century B.C. were the Stoics, the Epicureans and the Sceptics. The Stoics believed in all-embracing Divine Providence despite evidence to the contrary, such as occasions of natural disasters.

Chrysippus

Chrysippus was the most prominent and verbose of the Stoics. He argued that bedbugs had been created by a Benevolent Provider to stop people from over-sleeping. The Stoics made important developments in logical theory. This let them make arguments that had eluded Aristotle.

Epicuris (c. 342 - 270)

Epicuris was the founder of the Epicureans. They stressed that the experience of pleasure was the End ,but this consisted of the satisfaction of desires. They went on to say that they did not mean a lot of pleasure was a good thing. One should limit the number of desires a person had, just so you did not get left with too many unsatisfied desires. They were opposed to the view of philosophy that is a pursuit of the Ineffable and Unattainable Mystic Union with the Creator and Total Empathy with the Cosmos.

In the Letter to Menoeceus, the Epicureans stated, "By pleasure we mean the absence of mental and physical pain. It isn't a matter of boozing, orgiastic parties, or indulgence in women, small boys and fish." Why fish is not known but it is in the letter.

The Epicureans had their version of Atomic Theory, which was similar to Democritus, but in order to preserve Free Will, they held that at times the atoms swerved unpredictably, causing collisions.

They also believed that Gods existed but they did not give a damn about human affairs. The Gods had better things to do.

Pyrrho of Elis (c. 360 - 270)

Pyrrho was the founder of the Sceptics. They did not believe much of anything. Pyrrho did not write anything. Later Sceptics did write some books.

Timon wrote a book of lampoons called the Silli. Many titles of books were just as the author thought that they described the text.

Aenesidemus and **Sextus Empiricus** held in their writings the argument that no sensory report was trustworthy. It might be pleasurable, and therefore one could not be sure of anything. To support their view they had versions of the "Argument from Illusion" which Descartes later used.

Pyrrho's skepticism was so marked that friends had to stop him from walking off cliffs, flying under chariots,or jumping into rivers. His friends were good at saving him since he lived to a ripe old age. He supposedly visited the Indian Gymnosophists, or naked philosophers who were called that because they had naked seminars.

Minor Schools of Philosophy existed of which one was the Cynics. They were the masters of snide remarks

Philo (20 B.C. - 47 A.D) was a mystic Jew. He tried to fuse Platonic and Egyptian philosophies with Judaic teachings. He taught that the soul is immortal and the lost suffer eternal torment.

Crates was one of their leaders. He was known as the gatecrasher because he would burst into people's homes and insult them.

Diogenes was the most famous of the Cynics. He is reputed to have told Alexander the Great to get out of his light. He also scandalized people by eating, and making love in public.

The Cynics did not give a tinkers damn for anybody. They did not care what anybody thought of them. You could look upon them as models of a philosophic school or simply boorish oafs.

Athenagoras (127–190 A.D.) was a platonic philosopher and quasi Christian. He was the first in Christendom that we know of to use the term "immortal soul". He based this view not on scripture, but Greek philosophy.

Tertullian (160 - 240 A.D.)

Tertullian, Bishop of Carthage, advanced the idea of the soul's natural immortality and ceaseless torment for the lost souls.

Philosophy moved on in fits and starts in the Greco-Roman world under the unpredictable patrons whose attitudes varied considerably in regard to philosophers. Marcus Aurelius felt he was a philosopher. Nero killed the philosophers.

The influence of Christianity began to be involved in philosophy of sorts. Philosophy did not do well under the influence of Christianity.

Augustine

Augustine managed to become a saint. This happened despite his prodigal sexual promiscuity. He prayed where he asked God to "make me chaste, but not just yet". He anticipated Descartes Cogito: "I think, therefore I am."

Augustine also developed a theory concerning time, God stood apart from time, and he stated that God is not concerned with time.

Also during this period there also were the Neoplatonists. Some were Christian and some were not. Their names all seemed to begin with a "P". The Christians tried in vain to show that Plato had been a Christian.

Neoplatonists talked about abstract Things with capital letters, such as, One and Being which was used in a manner that no one could understand.

Plotinus, Porphyry and **Proclus** are three of the more notable of the Neoplatonists. Procius had some very interesting ideas about Causes.

Next came the Dark Ages. During this time philosophy was kept alive in the Arab countries and in the monasteries. European philosophy or what there was of it dealt mostly with theological concerns. Disputes between the monasteries whether God was one person in three or three people in God. Another topic was the exact nature of the Holy Spirit.

Chapter Four

The Dark Ages

For several hundred years
Jew, Arab and Christian lived together
where Religious intolerance was
not a fact of life.

During the Dark Ages one must take note of the city of Cordoba in Southern Spain. This part of Spain was under the control of and occupied by the Arabs who influenced philosophers during this time.

Maimonides

Maimonides was perhaps the greatest of the Jewish philosophers. His home was in the city of Cordoba.

Averroes

Averroes is also considered one of the greatest of the Arab philosophers. He also lived in Cordoba. Some scholars however, consider Avicenna to be the greatest Arab philosopher. For several hundred years Jew, Arab and Christian lived together in Cordoba; where Religious intolerance was not a fact of life.

Philosophy began to revive in Europe in the 11th Century. One of the philosophical saints among many was **Anselm**. He stated the Ontological Argument for the existence of God. This argument is remarkable for its implausibility, its longevity and its difficulty to refute it. Stated another way, it is to have something greater upon which nothing can exist. So if God does not exist, there would have to be a greater thing. We can accept this premise, so God must exist. Anselm said this argument came to him in a vision on July 13, 1087 - this is the first proposition in philosophy that can be dated.

Thomas Aquinas (1225-1274)

Aquinas was an important philosophical saint. He was

largely responsible for bringing the arguments of Aristotle back into the western philosophical mainstream. Aristotle had been ignored for several centuries. Some think it was because the western philosophers could not read and understand Greek.

Saint Thomas

Saint Thomas is the only philosopher officially recognized by the Catholic Church. He is noted for his stating the Five Ways of Proving the Existence of God. The first three ways were similar.

Saint Thomas also was noted for his two interesting arguments against incest. His first argument was that incest would make life more complex than it was and his Second argument that the bond between brother and sister would be all-powerful. He did not develop more on this second argument.

As for the rest of the Mediaeval Schoolmen, as they were known, the most important ones were the Franciscans. Specifically there were two men who were noted for this period.

Duns Scotus (1270 - 1308) was Irish and was noted 'Of reality the rarest veined unraveled,' but we know little of what that means.

William of Ockham (c 1290 - 1349)

William of Ockham is probably considered one of the greatest of the mediaeval logicians. He was best known for "Ockham's Razor" which was translated to: "Entities Should Not Be Multiplied Beyond Necessity".

Chapter Five

The Post Dark Age Of Philosophy

"I think, therefore I am."

Descartes

The post Dark Age of Philosophy starts with the Age of Greek Scepticism. Sceptical Epistemology formed the basis from which Descartes was able to rebuild a positive philosophy.

Rene Descartes (1596 -1650)

Descartes is considered the Father of Modern Philosophy. He invented the term Cogito while hiding in a stove to escape military service. He never married but he had an illegitimate daughter.

Descartes concluded that everything did not exist, except his own thoughts. He proceeded to rebuild a metaphysical bridge. He doubted the existence of God.

Philosophy split into British and Continental. The British were called Empiricists. This translates to building their philosophical systems on the basis of what could be felt, observed and experienced.

John Locke (1632 - 1704)

Locke thought that objects had two sorts of attributes.

The first one was primary qualities. This referred to extension, solidity, and number, which are held to be inseparable form and inherent in the objects themselves. The second was secondary qualities. This referred to color, taste, and smell, which seem to be in the objects themselves. These in fact are in the percipient.

Locke also said we have no innate ideas. He said the infant's mind at birth is a clean slate and all knowledge of the external world was learned from that which is experienced or extrapolated from that experience.

Locke had problems with personal identity. He had difficulty differentiating his mind from others. What is the content of the continuity of my personality? He wondered if he was the same person he was five years ago. He contended that all men were not necessarily persons. He contended that to be a person requires a certain level of intelligent self-consciousness; therefore, all persons, are not men.

George Berkeley (1865 - 1753)

Berkeley was Irish and a bishop. His ideas were more radical. He held that things exist only if perceived. His reasoning for believing this idea was only common sense. It is impossible to think of something being unperceived, for by thinking of something unperceived, by thinking of it you perceive it.

The philosophical refutation missed Berkeley's idea. People who subscribe to such views were called Idealists. For example **G. E. Moore** stated that Idealists believe that trains have wheels only while in the station. There are no wheels when they board the train because they can't be seen.

Hume (1711 - 1776)

Hume was the successor to this kind of Skepticism. Hume published his first book "*The Treatise of Human Nature*" in 1739. It was not well received. He then rewrote it under the title "*Enquiry into Human .Understanding*". This book was immediately a success.

Hume developed the idea that causes and effects are the names we give to events that have repeatedly been observed. Hume also thought that Free Will and Determinism could be made compatible.

Chapter Six

European Philosophy

*"Act only according to that maxim
by which you can at the same time will
that it should become a general law."*

Kant

Spinosa (1634 - 1677)

Spinosa was a Jewish lens-grinder from Amsterdam. With his Ethical system he created a set of formal deductions in geometry. He was a Determinist and believed in a logical necessity. His system was unsuitable to ethics.

Leibniz (1646 - 1716)

Leibniz was known as the carcature Pangloss in Voltaire's *Candide* This character thinks that everything is for the best in the best of all worlds.

Voltaire

Voltaire also wrote on logical and metaphysical subjects. Unfortunately these writings were not published during his lifetime. There was con-siderable difference between the quality of his private writings and the lack of quality in his public writings.

Rousseau

Rousseau and Diderot were the leading French philo-sophers of the eighteenth century. The two of them managed to be arrested or exiled or both.

Diderot

Diderot was noted for his originality, good sense, humanity and erotic prose. His writings were not read much by the British since the only one that was in English was "*La Religieuse*" (The Nun).

Marquis de Sade

Marquis de Sade was known for his brand of philosophy gone mad. His motto "If it feels good, do it" this is what he did and he was imprisoned for doing it. He wrote *Philosophie dans le Boudoin,* a mixture of socio-biological, political and moral philosophy intertwined with sado-masochistic sex. It is a wonder that his philosophy has been taken seriously.

The 19th century German philosophers must be mentioned. They were noted, mostly to the Germans.

Kant

Kant was probably the most noted of the German philosophers. He was famous for his "Categorical Imperative." He stated this as "Act only according to that maxim by which you can at the same time will that it should become a general law."

Hegel

Hegel had in common with lawyers, computer programmers, other German philosophers and preachers the talent of making what is simple fantastically complex. He used the word 'dialectic' to mean the interplay of opposing forces. This was important to the pre-history of Marxism.

Schopenhauer

Schopenhauer thought that philosophical terminology could be most impressive when used properly. Keep in mind that he was a German philosopher.

Nietzsche (1844-1900)

Nietzsche was a proto-fascist. He was also anti-Semitic, as was most of the population in the 19th century Prussia. He was of the opinion that God was dead or at the very least taking time off. He hated women.

Nietzsche advanced the Doctrine of Eternal Recurrence, where everything happens over and over. This subjects us to repetitious tedium remember the previous occurrences. Nietzsche was defiantly insane.

He wrote books with titles such as, *"Why Am I So Clever", "Why I Write Such Wonderful Books."*

Kierkegaard

Kierkegaard was one of the 19th century philosophers of note. He mostly rewrote other philosophers writings, but wrote little of nothing of his own.

Henri Bergson

Bergson was a Vitalist, believing the difference between animate and inanimates, the presence in the animate of ***a*** mysterious force "Elan Vital", or Life Force. This Life Force he postulated leaves the body during adolescence. He also wrote a book about laughter

The distinctive contribution to philosophy at this time was Pragmatism, which questioned the meaning that truth or falsity is not absolute but matters of convention. **William James** and **John Dewey** held this view.

Chapter Seven

20th Century Philosophers

Infinite Regress

The philosophical equivalent of banging your head against a brick wall

Ethics

Much of the study of the 20th century philosophers had to do with broad based Ethics and covered many professions.

A meta-language is the structure of another language. With the meta-language there is the opportunity of an *Infinite Regress,* which is the philosophical equivalent of banging your head against a brick wall.

Alfred Tarski

Tarski was a logician that held that by positing an infinite hierarchy of languages could we fully explicate the function of truth as it functions in ordinary language. The reasons for this are complex and difficult and can be mastered only after years of study. This was the reasoning of Tarski.

Donald Davidson

was an American philosopher that agreed with and added to the reasoning of Tarski. His writings were even more difficult to understand. He adapted philosophy and languages to a *"Theory of Meaning"*. He started as a theoretical psychologist, but switched to philosophy.

There are basically two types of ethical theory: Consequentualist or Dentologist. The former holds that an action's moral quality is determined solely by its results. The latter holds that there some things that one ought to do, and others that one should refrain from, no matter what the actual or probable results of that action.

Bentham and **Mill** had a version of consequentialism they named utilitarianism. This variety of ethic holds that one should act to produce the greatest good for the greatest number. How good is to be defined and by whom, and what to do in the case of incompatible goods, and whether the total number of people involved matters. There is difficulty in interpreting these problems.

Deontologists are more difficult to understand. In keeping with arguments that are well covered with the Subjectivist-Objectivist debates. At question, are morals discovered or are they simple matters of convenience being created in an arbitrary way to make social or anti-social activity possible.

This brings us to cultural relativism where people who adopt this position hold that no society has the right to say what is right or wrong about any other society.

Dick Hare has stated that he was never able to understand the difference between the two positions and that he had never met anyone else who could understand the difference.

A very useful technique is to pretend that the subject is very obvious when in fact it is obscure. **Wittgenstein** used this technique at times. However **Dick Hare** was the master of this technique. He once claimed he did not understand the word 'it'.

For the classical versions of Deontologies one would look to **Kant** with his famous categorical imperative He has developed several different formulations in his writings.

This is generally expressed as: "Act only according to the maxim by which at the same time will that it should become a general law." This statement is interpreted to mean you should only do things you would not mind if everyone else did it.

Advocates of deontologies are conservative churches, conservative politicians and some religious colleges.

Philosophers tend to think less in terms of duties than of rights, and to work at creating Rights Theories. An example is that you are deserving something or be allowed to get away with.

The most important principle in Rights Theories is that you or I have a right only if everyone else has that right. Rights have extremely good value. Keep in mind that most people that think they have rights usually hold inconsistent beliefs about them. For example, everyone has a right to life; however the oppressed have a right to kill tyrants. If we are in a war it is all right to kill the enemy.

There can be a distinction between Act-centered and Agent-centered theories of ethics. This is about whether what really matters in morals are what sorts of the things we do, or the kind of people we are.

Another set of ethical questions is the New Moral Problem. These are ethical questions in the field of medicine such as test tube babies, cloning, stem cell research and fetal experimentation.

Euthanasia could be included here but it has been practiced in one form or another almost from the beginning of the human race. You can make a distinction about active and passive euthanasia. Active is actually killing ,passive is merely letting a person die. A doctor that refuses treatment intends for the patient to die. He will be responsible for the patient's death as surely as if he shot the patient with a gun. The recent court cases have brought euthanasia more to the forefront in determining ethical positions.

The question of a woman's right to have an abortion has been settled by the courts although it is still being argued by some sectors of the population. If a woman has the right to choose an abortion, does this extend to the right to kill. The question is does a fetus have rights. The courts have indirectly ruled on this. A fetus does not have rights.

Logic

Another important branch of philosophical enquiry is Logic. Logic is very difficult to comprehend.

Gottlob Frege

Frege is the person that took logic beyond Aristotle's position. He was the leader in the development of mathematical logic.

Russell and **Whitehead** were also involved in the development of mathematical logic. This new logic could do more computations than before.

Technically, it can deal with the logic of relations between things. It cannot deal with human thought processes. This is much harder to understand.

In the 20th century we also developed alternative logics. This is known as deviant logics. These studies increase the number of truth-values and question the validity of certain traditional logical laws. An example is the Law of the excluded middle which says that everything is something or it isn't.

In addition to Formal Logic we also have Philosophical Logic. This is a large and confusing area of logic. One of the central concerns is 'Theory of Meaning', which many philosophers reject.

Epistemology

Currently epistemology is titled cognitive science. This is a mixture of logic, linguistics, psychology and computer science. It is concerned with modeling reasoning, human as well as artificial intelligence. As far as humans are concerned it makes much use of deviant logics. Cognitive science tends to be complicated and difficult to understand.

Philosophy of Religion

Some of the questions that are debated among persons that consider themselves a philosopher of religion are does God exist, and if so what He thinks and what is He doing. The problem of evil and whether God loves the world and us are other questions.

Lactantius gave us some choice arguments. Given that there is evil and bearing in mind the supposed attributes of God, then either: God knows about it, cares about it, but cannot do anything about it. He would care about it, could do something about it, but doesn't know about it. He knows about it, could do something about it, but doesn't care about it.

Philosophy of Science

There can be a philosophy of almost anything. The philosophy of science is one example. Questions that might be asked are as follows. How do scientists develop theories? What is the relationship between theory and evidence? What is the so-called experimental method? How does one theory become more accepted than another competing theory?

The philosophy of science has enjoyed a booming business in the last one hundred years, because it gives scientific philosophers the feeling that what they are doing is related to something they feel is of importance, even if it is not important.

Karl Popper

Popper held the view that theories can never be verified. They can only be proven false. He stated that no amount of empirical data can show beyond a doubt that the world will always continue to happen in the same way. According to Popper, proper theories are made up of exceptionless generalizations, universally quantified. Stated another way they consist of sentences in the form of “All somethings are somethings else”.

Other people have offered different meanings. **T. S. Kuhn** talks about scientific revolutions that involve what he called paradigm shifts. Broadly he means that people decide to stop looking at the world in one way and decide to start looking at it in another way.

Hilary Putnam

Putnam stated that if Popper were right, no theory is falsifiable. Putnam had the habit of changing his subtle and complex views just when his contemporary philosophers were beginning to understand one of his theories.

Current day English and American philosophers deny that they belong to any particular school of philosophers. However they belong to the 'Analytic Philosophers' group, even though most of them never analyzed anything.

British Philosophy

British Philosophers of importance were **Bertrand Russell** and **G. E. Moore.**

Bertrand Russell

Russell made his reputation by writing the book *Principia Mathematic"* in cooperation with **A. N. Whitehead.** This book was a very detailed description of formal symbolic logic.

G. E. Moore

Moore produced his *Principia Ethic"* book. On the subject of moral philosophy, he held that "good" was indefinable, but was the name of a non-natural quality.

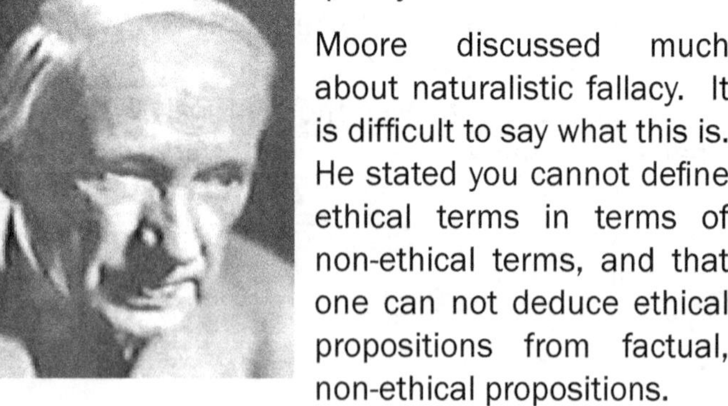

Moore discussed much about naturalistic fallacy. It is difficult to say what this is. He stated you cannot define ethical terms in terms of non-ethical terms, and that one can not deduce ethical propositions from factual, non-ethical propositions.

Structuralism and post-structuralism have become known as post-modernism. It started as linguistic philosophy and rapidly became involved with anthropology. characteristic of the literature is distrust of academic disciplines and its jargon.

Leading supporters were **Roland Barthes** in the field of literary criticism; **Michel Foucalt** in the fields of history, sociology, and sex; and **Jacques Derrida** in the fields of language, literary criticism and rhetoric.

Derrida in the field of rhetoric used a method known as deconstructionism. He showed that any literary effort of necessity has within it fatal contradictions which undermine the literary effort. He claimed that what is said is meaningless, and he claimed derisively, that whatever it is, it is not philosophy. He remarked cautiously that it should not be dismissed out of hand.

American Philosophy

An important man in American Philosophy was

Willard van Orman Quine

Quine's book of note was *From A Logical Point of View.*

Another individual was **Saul Kripke**

Kripke's book in philosophical logic, *Naming and Necessity*, concerns proper names, sense and reference, and possible words.

Alvin Plantinga

Plantinga was a Viennese positivist that ended up in America. He was a modal logician and a philosopher of religion. He took the position of being fashionable and skilled in the art of logic.

John Rawls was an important American philosopher who wrote *A Theory of Justice*. He held that justice could be placed within two principles. Everyone should have the same freedom as everyone else, and

given that constraint, as much freedom as possible. That economic inequalities between people are only justified if the worst off are in fact better off under this arrangement than they would be under any more equal one. This type of logic is also known as the Theory of Distributive Justice.

Continental Philosophy

Logical positivism had its supporters more closely related to Anglo-Saxon tradition. Many of its supporters fled Europe and the Hitler regime and immigrated to America. **Carnap and Hempel** were very influential, particularly in the philosophy of science.

Rudolph Carnap

Carl Hempel

The English philosopher

A. J. Ayer

Ayer became well known for his first book, *Truth and Logic*.

Continental philosophy basically was either French or German. Existentialism had followers in both countries

French Philosophy

Sartre

The principle French follower was **Sartre**. He combined his philosophy with active Marxist politics. He coined the slogan "existence precedes essence". Which means we should be less concerned with the types of things than with the fact that they exist.

Existentialists deny classifications. They insist on the autonomy of the individual. French existentialism had strong literary connections. Their literature tended to be of a violent or anti-social nature.

German Philosophy

The Germans were a different group of philosophers. **Jaspers** and **Heidegger** had no pretensions for literary excellence but were explicit in their writings. They were systematic in their writings about Phenomenology. Which was the attempt to penetrate the surface of appearances of things to know what our conscious apprehension of them.

Existentialism to the Germans carried no religious connotations. Sartre was an atheist. Camus was a Christian. Heidegger was a Nazi.

Jaspers

Heidegger

Books on philosophy were thought to generally have three elements: Language, Truth and Logic. An example would be Heideggers book on "Being and Nothingness".

The Germans despised the Anglo-Saxon analytic philosophers for not being sufficiently accurate in their writings. This included the American philosophers.

Philosophical Dilemmas

These philosophical dilemmas are excellent topics for group discussion. With what you now know about philosophy you should be able to take the pro or con position. It will be left to you to develop your position.

- **Is Faith an Answer?**
- **Is Liberty the Highest Social Value?**
- **Is Equality the Highest Social Value?**
- **Is Pleasure The Only Value?**
- **Can Happiness Be Defined?**
- **Is Morality Relative?**
- **Is Happiness The Standard of Morality?**
- **Should Doctors Ever End People's Lives?**
- **Are We Free?**
- **Are We Responsible For Our Actions?**
- **Is The Mind Nothing But The Brain?**
- **Are We Always Selfish?**
- **Can Experience be the Source of Knowledge?**
- **Does Truth Exist?**
- **Does the end always justify the means?**

Philosophers Covered In This Text

SEVENTH CENTURY B.C. TO THE END OF THE PRE-SOCRATICS

Herodotus	Thales of Miletus
Anaximander	Anaximenes
Heraclitus	Pythagorus
Empedocies	Paramenides
Melissus	Zeno
Democetus	Leucippus

SOCRATES TO ARISTOTLE

Socrates	Plato

ARISTOTLE TO THE DARK AGES

Aristotle	Speusippus
Chrysippus	Epicuris
Pyrrho of Elis	Aenesidemus
Sextus Empiricus	Philo
Crates	Diogenes
Athenagoras	Tertullian
Augustine	Plotinus
Porphyry	Procius

THE DARK AGES

Maimonides	Averroes
Avicenna	Ansiem
Thomas Aquinas	Duns Scotus
William of Ockham	

POST DARK AGES

Rene Descartes	John Locke
George Berkley	Hume

EUROPEAN PHILOSOPHY

Spinoza	Leibniz
Voltaire	Rousseau
Diderot	Marquis de Sade
Kant	Hegel
Schopenhauer	Nietzsche
Henri Kienikegaard	William James
John Dewey	

20TH CENTURY PHILOSOPHERS

Ethics

Alfred Tarski	Donald Davidson
Bentham	Mill
Dick Hare	

Logic

Wittgenstein	Gottlob Frenge
Russell	Lactantius
Karl Popper	T. S. Kuhn
Hilary Putnam	

British Philosophers

Bertrand Russell	G. E. Moore
A. N. Whitehead	

Post-Modernism Philosophers

Roland Barthes	Michael Foucalt
Jacques Derrida	

American Philosophers

Willard van Orman Quine	Saul Kripke
Alvin Plantings	John Rawls

Continental Philosophers

Rudolph Carnap	Carl Hempel
A. J. Ayer	Sartre
Jaspers	Heidegger

Sources

Plato. *Symposium; The Dialogues of Plato,* Volume 2 Second Edition

Gorgias: The Dialogues of Plato, Volume 2, Second Edition Translated into English with analysis and introduction by Benjamin Jowett, Oxford: Clarenden Press, 1875

Republic: The Republic of Plato Translated in English by Benjamin Jowett. Oxford: Clarenden Press, 1888

Aristotle. *The Politics and Economics of Aristotle.* Translated, with notes and analysis, by Edward Walford London: Bell & Daldy, 1871

Machiavelli, Niccolo. *The Prince.* Translated by Paul Rahe.

Adler, Mortimer J. Ten Philosophical Mistakes. London, Collier Macmillion, 1902

Augustine. *Confessions.* Translated by Edward B, Pusey. New York, P. F. Collier & Son, 1909

Hagel, Georg Wilhelm Fredrich. *Lectures on the Philosophy of Religion, Volume 1.* Translated by E. B. Spiers and J. Burton Sanderson. London, Kegan Paul, Trench, Tubner, 1895

G. S. Kirk, J. E. Raven, M. Schofield *The Presocratic Philosophers. Presocratic Philosophers,* 2nd Edition., Cambridge, Cambridge University Press, 1983

Pythagoras. *Pythagoras, A Life. P. Gorman, London, Routledge and Kegan Paul, 1979*

Aristotle. Aristotle's Theory of Poetry and Fine Art. Translated by S. H. Bucher, London, Macmillan, 1907

Kant. *Prolegomenea to any Future Metaphysics.* Traslated by Lawrence G. Horshburgh

Hume, David. A Treatise of Human Nature, etc. Abstract of the Book, 1740

Hintikka, Jaakko. *Principles of Philosophical Reasoning.* Edited by James H. Ferzer, Totowa, N.J., Rowman and Allanheld, 1984

Husserl, Edmund. *The Crises of European Sciences.* Translated by David Carr, Chicago, Northwestern University Press, 1970

Socrates. *Essays on the Philosophies of Socrates.* H. H. Benson, Oxford, Oxford University Press, 1922

*Augustine. Augustine.*C. Kirwan, London, Routledge, 1989

Maimonides. *Maimonides on Human Perfection.* M. Kellner, Atlanta, *GA,* Scholars Press, 1990

St. Thomas Aquinas. *The Thought of Tomas Aquinas.* B. Davies, Oxford, Clarenden Press, 1992

*William of Ockham. William of Ockham.*M. M. Adams, Notre Dame, IN, Notre Dame Press, 1987

Francis Bacon. *Essays.*Everyman Library, 58 Essays, London, Dent 1968

Galileo. *Galileo Studies.*Translated by J. Mepham, Hassocks, Sussex, Harvestor Press, 1978

Thomas Hobbs. Leviathan. (1651) English Works, Volume III, Edited by C. B. Macpherson, Harmondsworth, Penguin 1951

Rene Descartes. *The Essential Descartes.* Edited by Margaret D. Wilson, New York, Mentor Books, 1969

John Locke. *Locke.* Past Masters Series, Oxford, Oxford University Press, 1984

Empiricists. *Critical Essays on Locke, Berkley, and Hume.* Edited by M. Atherton, Lanham, Rowman and Littlefield, 1999

Kant. *Critique of Pure Reason.* (1781) Translated by N. Kemp Smith, London, Macmillan, 1929, Corrected 1933

Stanley Rosen, Editor. *Essential Readings from Plato to Kant.* The Philosophers Handbook, NewYork, Random House, 2000

Jeremy Bentham. *Fragment on Government.* (1776) Cambridge, Cambridge University Press, 1988

John Dewey. *Dewey's Metaphysics.*R. Boisvert, New York, Fordham University Press, 1988

Bertrand Russell. *The Problems With Philosophy.* (1912) Oxford, Oxford University Press, 1973

Diane Collison and Kathryn Plant. *Fifty Major Philosophers* London, Routledge, 1988

Karl Raimund Popper. *The Open Society and Its Enemies.* London, Routledge, 1945

Willard van Orman Quine. *Word and Object.* Cambridge, MA, MIT Press, 1960

Jean-Paul Sartre. *The Psychology of the Imagination.* (1940) London, Methuen, 1983

Philosophy and Philosophers
Great Thinkers and the Way They Think

ISBN 978-1-935434-03-0

GlobalEdAdvancePress™
37321-7635 USA
www.GlobalEdAdvance.org

www.ingramcontent.com/pod-product-compliance
Lightning Source LLC
LaVergne TN
LVHW010106110826
845155LV00028B/510